How to boil an egg

How to boil an egg

A fresh look at sustainable energy for everyone

Ad van Wijk

IOS Press

ISBN 978-1-60750-989-9 (print)
ISBN 978-1-60750-990-5 (online)
doi: 10.3233/978-1-60750-989-9-i

Publisher
IOS Press BV
Nieuwe Hemweg 6b
1013 BG Amsterdam
The Netherlands
tel: +31-20-688 3355
fax: +31-20-687 0019
email: info@iospress.nl
www.iospress.nl

Co-publisher: mauritsgroen•mgmc (www.mgmc.nl)

Author: Ad van Wijk
Translation: Milja Fenger
Editing: Ed Mears, Rolf de Vos
With contributions from David de Jager, Gijs de Reeper and others
Illustrations: SYB (Sybren Kuiper)
Graphic design/lay out: Tekst in Vorm

PRINTED IN THE NETHERLANDS

Contents

The energy crisis does not exist

I have a vision. In that vision humanity is doing all that lays within her power to halt climate change. In that vision everyone in the entire world can fulfill their own energy needs from clean, inexhaustible energy sources. Wars are no longer fought over energy.

This vision may yet become a reality. It is, in fact, possible to fulfill our abundant 'energy desires' in a clean manner, and often with surprising ease. Then what have we been doing for the last few decennia? Sleeping? Certainly what we have done for years is to look in the wrong places for solutions to our problem. This booklet shows energy in a different light. And hopefully that will bring us new ideas that are good for the climate, the environment, and your wallet.

Sources of inspiration

In 1983 I was a physicist at the University of Utrecht in the Netherlands. I was researching the potential of wind energy. Ten years before that the Middle East had halted the oil supply to several countries (including the Netherlands). This brought to light our energy dependence and vulnerability. At the same time the Club of Rome warned us of the finite nature of several natural resources, like the fossil fuels we are apparently so reliant on. I started thinking about how it might be possible to build a sustainable future without exhaustible energy sources such as oil, coal, natural gas, and uranium; and, without sacrificing everyday comforts.

I was born in 1956 and grew up on a small farm in the centre of The Netherlands. We had a mere twenty cows, a few pigs and a number of chickens. We cultivated our own vegetables and potatoes. Almost everything our family needed came from the land that belonged to the farm. We didn't have much. Sometimes we had to live from hand to mouth.

We slept amongst the cows because their body heat kept us warm. We gathered firewood so we could cook food and heat the kitchen. The sun dried the grass turning it to hay, so that in the winter we could feed the cows. The wind and the sun dried our clothes after being washed by hand. Horsepower, in the shape of an actual horse, plowed the field and brought us to church on Sundays. Only at the end of the sixties could we afford to buy a tractor and a car and let our old equestrian friend retire. And, we finally had electricity, a TV, and a shower.

These and other technological advances made life easier and more comfortable. For poor people like us, this higher standard of living had a major impact. I don't cherish any nostalgic longing for the 'innocent time' when the power plug did not yet

exist. But, my youth on the farm has given me an awareness of the energy around us which is freely available: wind, sun, organic waste and, of course, the body heat of cows. Even though these energy sources are found nearly everywhere, they are not always easy to turn into something useful and practical. For how can you dry your clothes in the Northern-European damp winter air when there is no sun?

What really matters is: How can we convert primary energy into reliable and available power? Oil, gas and coal (and sun, wind and water as well) require energy to be converted and stored before use. You have to do something with any of these primary energy sources in order to keep your beer pleasantly cool, light your room or heat your relaxing bath. Mankind, in the last few centuries, has come up with several intelligent systems to do just that.

But these systems often function far from optimally. Take as an example the energy-rich cow dung currently falling on our fields as a waste product. To be able to make use of the dung, it first has to be fermented to create gas. Then it has to be burned in a turbine where a generator makes electricity that will feed your TV. Is there no way to do this more cleverly? What is the shortest path from cow dung to your TV?

During the 19th century Industrial Revolution we started to learn how to fulfill most of our energy needs by using fossil fuels. The discovery of the great oil and gas fields in the 50s and 60s gave a definite push towards an economy and a source of worldwide wealth fueled by fossil energy. The use of oil, gas and coal has been so imbedded in our daily lives that we now fail to see other cleaner energy sources, ones that are easy to find close to home. These kinds of clean resources will be explored in this booklet. A fresh perspective in the way we think about energy is needed. It will soon become apparent that scarcity is not a problem. To the contrary: the possibilities are endless. It is the creation of a sustainable energy supply that proves to be difficult. But an energy crisis does not exist. I will prove it to you.

Energy production in a greenhouse

Like many farmers at the time, on our small farm we cultivated our vegetables exclusively in the open field. We were subjected to the whims of the Dutch weather. Temperature, solar irradiation, wind, hail, insects and floods determined the yield, and these factors could not be controlled.

Greenhouses first became common in Dutch agriculture around half a century ago. These were simple low-lying, horizontal plates of glass that saved the heat from the morning sun, allowing seeds to develop into seedlings more quickly.

In Holland the interest in greenhouses received a major boost after the 1959 discovery of a gigantic natural gas field near the village of Slochteren, in the north of the Netherlands. This was the world's second largest natural gas discovery at that time. The gas made it possible to heat the greenhouses in winter. The yields multiplied and more exotic fruits and vegetables such as tomatoes, bell peppers and strawberries could be harvested year round. Greenhouses also helped to turn the Netherlands into one of the world's largest exporters of flowers. Currently around

10% of the yearly production of natural gas is used in Holland's horticulture industry, and it is worth many billions of euros.

Essentially a glass greenhouse does little else than collect solar energy. Modern greenhouses can collect solar heat so effectively that in the summer the windows are often opened to let excess heat escape. But when you let the warmth out, pests such as insects and fungi enter. Opening the windows also lets the CO_2 escape and CO_2 is an effective means of making the plants grow more. In contrast, the Dutch winter sun is too weak to heat the greenhouses and farmers are forced to rely on natural gas to heat them.

At first glance a greenhouse owner is faced with two problems: in the summer there is too much energy and in the winter there is too little. These problems are mirror images. Is it possible to let one solve the other? My answer: Yes, indeed.

The 'Closed Greenhouse' has solved this problem by storing its excess summer heat in underground water. The heat is transported through a pipeline some 50 to 100 meters into the ground under the greenhouse, in natural underground water containers called 'aquifers'. These aquifers are subterranean pockets of sand, saturated with water and surrounded by clay. One key feature of aquifers is that water is kept at a constant temperature over long periods of time. In the winter this process works just the opposite of the way it works in the summer. The relatively warm water is transported back up, heating the greenhouse.

Aquifer systems require little advanced technology. The greenhouse is used as a solar collector while an aquifer nearby is a used to store (and later "recycle") the accumulated heat. A simple pipeline and a pumping system provide the final touch. The advantages of a closed greenhouse are not limited to solving the heating issue and saving a lot of natural gas. For instance, by keeping the windows shut the irrigation water stays in the greenhouse. Farmers can expect to cut their water bill in half. Second, insects and other pests cannot get to the plants. The use of pesticides can be eliminated, and the produce may be labelled 'organic'. Finally, since higher levels of carbon dioxide are maintained, plants produce about 25% more.

'Tweaking' an ordinary greenhouse in this way is very rewarding for a farmer. In contrast with the typical farmer in the 1950s, a farmer today is able to manipulate the circumstances and growing seasons at will. This way he can choose to grow and harvest red roses right before Valentine's Day.

In practice the Closed Greenhouse often produces such vast amounts of surplus heat in summer. This surplus can be used in the winter to heat surrounding buildings, other greenhouses, residential buildings, office buildings, or even a swimming pool. In effect, the greenhouse becomes an energy producer!

At present, the Closed Greenhouse is becoming a real revolution in greenhouse horticulture. And this is only one example of how this technology can be used. About 25% of all our energy consumption in the world is used for heating and cooling our buildings. Expanding the Closed Greenhouse methodology may profit us immensely.

Four Laws

There are many roads that can lead to a truly sustainable energy supply. Roads that can be walked by a variety of groups with creativity, courage, and an entrepreneurial spirit. To inspire these groups, in this booklet I will explain some basic rules that are the foundation for any sustainable energy supply system. I call these the Four Laws.

The changes suggested in the Four Laws may be held back in certain situations by other laws and concerns. But in each of these laws, without a doubt, lays the possibility of a sustainable solution.

98%

From the extraction of oil, coal and gas, all the way through to delivery of light, heat and power, we squander a lot of energy. If you define the 'efficiency' of the energy supply as the quantity of the energy that is practically used divided by the energy dug out of the ground, the overall efficiency would be somewhere around 10%. This means that around 90% of this costly fossil energy is lost along the way.

The situation is even worse when we investigate what happens at the very end of the energy supply chain. That 'end' contains all those office buildings that are lit up and heated during the night, when no one is at work. That end is where a single car moves a ton of steel over the road, when the body it is transporting weighs less than a hundred kilos. Also at that end is the egg that is boiled in litres of water that are destined to disappear down the drain.

If we take into account all of this unnecessary waste at the end of the chain it becomes apparent that we do not use much more than 2% of all the energy extracted from the earth in the form of fossil fuels. We do not need that boiled water: we need a boiled egg. We do not need to move our cars: we want to move ourselves. And how about the lamps lighting the 11th floor of an empty office building? Let's not even think about that.

Right now we are throwing away 98% of our energy, just like that! We can do it differently, but let's not start at the beginning. Let's start at the end, where our collective needs and desires are given a material reality. In this world turned upside-down we will start at the book we want to read, the boiled egg we need to eat, or the desire to move from point A to point B. Forget how it is done for the moment. Let's leave the old behind and design a new system as if the old one no longer exists.

One hour

The sun is our everyday energy source. Worshipped in ancient times, we seem to have forgotten its main features. The power of the sun lies at the heart of virtually every form of energy and is found in many guises. First of all there is direct sunlight and the heat stored in the atmosphere, water, and soil. Wind energy derives from the sun, too, as does the power embodied in water cycles and waves.

Wood and other forms of biomass are no more than stored solar energy. When that storage has lasted for millions of years, we call the resultant product coal, oil or natural gas. Tidal energy is an exception, because it derives from the gravitational force of the Moon. In addition to these entirely renewable sources, the Earth itself supplies us with energy in the form of geothermal heat from the deeper layers of its crust.

The conclusions we come to are truly astonishing when computing the quantity of solar energy that is available. Every hour of the day enough sunlight reaches the Earth to cover humanity's entire current annual energy needs! In other words, every hour of every day the Earth receives around 10,000 times as much energy as its inhabitants are currently using.

Everywhere

The accessibility and availability of fossil fuels and the centralized energy systems built around them have destroyed our sense of 'the sources of old', the sources of energy that I experienced as a child on the farm. Although we are practically sitting on top of endless alternative possibilities, we fail to recognize them. Sun and wind are everywhere. Water is almost everywhere. And these are only a few of the energy sources waiting to be explored.

This bare fact has become the basic principle behind the way I look at energy supply systems. When you begin to search, you will find large quantities of free and renewable energy. The golden rule: look around you. Energy is everywhere.

Everyone

Fossil fuels and their derived products such as petroleum and electricity are not equally available to everyone. For a variety of reasons, two to three billion people today still lack this kind of commercial energy.

The sun rises for free and the wind blows for everyone, rich or poor. Much the same can be said about plants. The defining characteristics of most forms of sustainable energy are that their presence is practically guaranteed and that “harvesting” them does not deplete the supply.

If we want to extract usable energy from sun, wind or water then we will have to accept that this costs money. But it is also important to realize that it can also be done cheaply and almost anywhere in the world. Once the “energy yielding machines” have been bought or created, the energy production itself will hardly cost anything because the respective source is supplying primary energy for free. This is fundamentally different from energy production with fossil fuels, where all the resources and derivative products have to be paid for as long as production continues.

There are also certain downsides to sustainable energy sources. Firstly, renewable energy sources need a lot of space; for example, to harvest the energy from the sun, the wind or plants. This is very different from the concentrated energy in oil, coal and gas. Secondly, the renewable energy sources are often inconsistent. The sun only shines during the day – and with varying strength - and the wind occasionally does not blow at all.

Fossil fuels provide a constant and easy-to-regulate supply of energy, while most sustainable sources are less controllable. However, the abundant and almost cost-free availability of clean sustainable energy stands in stark contrast with the finite supplies of unclean fossil fuels, and the large expense required to obtain them.

Thinking differently

At the farm we had about a hundred chickens, producing about a hundred eggs each day. Most of these were sold, and we ate them ourselves practically every day. We had boiled eggs at breakfast, on bread, and in salads. We boiled our eggs on a stove fired by wood. So one could say that in my childhood we used sustainable energy to cook our eggs. But it was not very convenient, because we had to find the wood on our land or cut down a tree. I can tell you that I as a child I was less than happy with my daily chore of going to the forest and filling my basket with wood for the stove. I thought the gas oven was a piece of much desired progress. Naturally I myself do not have to send my children to the woods to fill a basket with wood. I would not want to do this and it is not necessary since we live in a prosperous country. What we do have to do is to develop a sustainable system that uses exactly the amount of energy that is needed to boil our egg.

Finding these solutions requires primarily that we start thinking differently about energy. We need a fresh outlook, if not a true shift in paradigm. Only after that has taken place can we develop an energy system that is totally different from the current one. Once we throw the ballast off the ship, what remains is a sense of surprise about how simple, cheap, and clean the solutions can be.

98%

Energy is in the news. Will we build more coal-fired power plants? Will we build wind turbines onshore or offshore? Is nuclear energy our future? The "problems" associated with these energy sources are usually central to any media discussion about them. The concerns expressed are usually about the varying demands on the environment, the depletion of resources, the costs to extract and convert the resources into electricity or petrol, and the geopolitical tensions that extracting oil or gas have caused at times.

The media attention to sustainable energy is becoming ever more apparent. This attention is both positive and negative. But in both cases the alternative is taken seriously, which in itself is a positive development. However, let us not confuse sustainable and fossil energy. We will not reach our goals by simply replacing a coal-fired power plant with a wind farm.

Here is how the current energy supply system works. First, a very large amount of resources (oil, coal, gas) is extracted at a limited number of locations on this planet and then transported to locations closer to the consumer. Second, at this new location a conversion takes place at refineries, power stations, and water boilers. In our current system the energy meter or the petrol pump is seen as the end station for energy.

In the conversion of fossil fuels to electricity large quantities of useful energy are lost. In a modern coal-fired power plant an energy efficiency of around 50% can be achieved. In a gas-fired power plant this efficiency is at most 60%. This means that a mere half of the energy stored in oil, gas, or coal actually reaches the electrical grid. The other half is lost as heat in water used for cooling or dissipates into the air. When we take into account the losses and energy requirements of extracting and transport, only 40% is eventually useable while 60% is lost before reaching the electricity outlet in your house.

Right now many thousands of energy-experts in energy companies, governments, and universities are working hard to improve the efficiency of each step in the process by a few percentage points. Their dedication deserves our appreciation, as does the expertise of these specialists. However, to take a far-reaching step toward sustainability, these experts may be looking in the wrong place.

The largest energy efficiency savings can be achieved at the other side of the electricity meter or petrol pump, at the precise place where energy is transformed into something we really want: a warm house, a glass of coke, some light to read a book by. The energy efficiency of an ordinary light bulb is no more than 2%, so 98% of the electricity used to illuminate its surroundings is lost in unwanted heat, not light. That is the reason we are now replacing conventional bulbs with LED bulbs. Their ratio of energy use versus energy loss is far better.

Finally it is not the energy itself that is important; it is the energy's function. The difference between the two terms is more than just semantics. People want to get their children to school, use their computer, watch TV, have a hot shower, or eat a warm meal, and it is irrelevant whether these 'functions' of energy are fulfilled by coal, solar energy or cow's dung.

So let's examine how the energy required for a particular function might best be supplied, instead of simply relying on the nearest power point or gas pipe. Because the largest losses occur at the end of the energy supply chain it is there we should start looking for savings.

How to boil an egg

Many people enjoy a boiled egg for breakfast. Let us consider the routines we go through in order to get one on our plate. Every year mankind consumes around 1,200 billion eggs.

We fill a pan with water and place it on a stove heated by gas or electricity. Or, in large parts of the world they use coal, wood or oil. The heat of the stove makes the water bubble and steam within a few minutes. The eggs are placed in the hot water: five minutes for a soft-boiled one, around ten minutes for a hard-boiled one. The eggs are taken out of the pan, the water is poured down the drain into the sewers, and the pan is rinsed and washed. Now finally: bon appétit!

What can we conclude about this simple process from an energy point of view? That it is not great. The energy efficiency of our morning routine is extremely low. At most 10%, more likely 5%, of the gas or electricity we use is used effectively. The rest disappears into the atmosphere of the kitchen and into the sewer along with the now useless heated water.

How can we improve on these low overall energy efficiencies plaguing our daily lives? In principle there are two methods. Either we improve on the efficiency of the process of cooking itself, or we reuse the heat we are currently throwing away. For the first option we have several technologies at hand. We could improve the quality of the stove, but realistically we will not achieve major gains by doing this. A simple, but greater improvement would be to reduce the amount of water used to boil our eggs. Heating a liter of water obviously uses twice the amount of gas as is needed

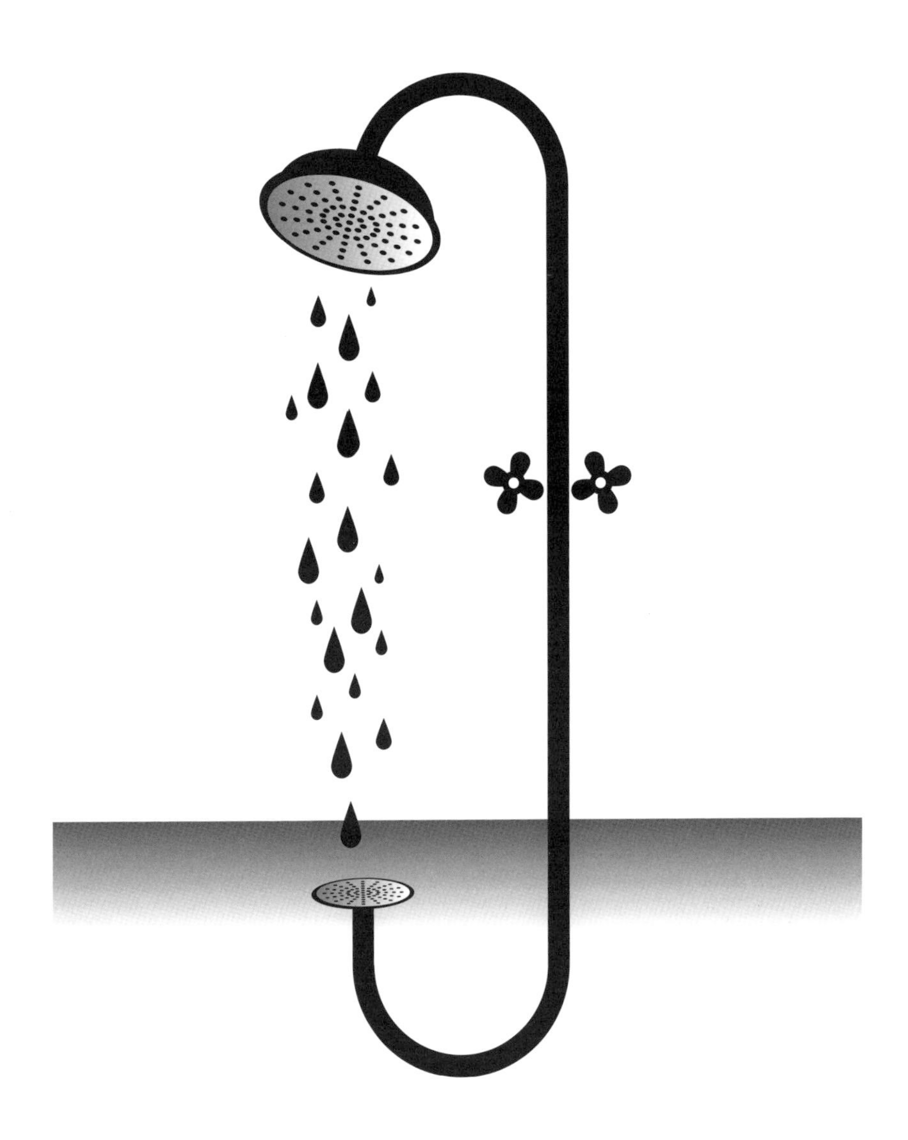

to heat half a liter of water. With a high-pressure cooker, for example, you use less water and need less time to boil that egg.

In order to be consistent in my line of reasoning, I have to conclude that it would be best to boil our egg without any water at all. A microwave offers us that possibility, because it cooks without water. However, as many of us know, an egg and a microwave together equal disastrous consequences. The egg explodes because the microwave radiation affects various parts of the egg in very different ways.

I will skip other cooking possibilities that require additional research or maybe a revolution in our kitchen that we are not yet prepared for. Instead let's focus on reusing the energy that is thrown away. Currently many modern kitchens are equipped with a small built-in boiler and a small reservoir under the sink that rapidly produces hot water from the tap. With a simple $25 device called a heat exchanger – which in essence is no more than an extra small pipe in the sink – the heat from the discarded water can be rerouted to the close-in boiler. This means that less electricity will be needed to heat water that is used later on.

Taking a shower

For a considerable portion of the world's population, taking a shower has become an everyday habit. Like many parents, I tell my children not to take a long shower, for environmental or financial reasons, or just because I don't want to wait too long to take my shower. This is helpful, but it is not the point I am trying to make. Let's have a look at the energy efficiency of this everyday activity.

Household boilers generally heat water to about 60 degrees Celsius (140 degrees Fahrenheit). For a comfortable shower this hot water is mixed in the tap with cold water to around 35° C. The warm water sprays out of the showerhead, is in contact with your body for a split second, and then disappears down the drain at a temperature of around 30° C.

It's not hard to imagine that some 50% of the natural gas energy has already been wasted from the time the hot water is produced to the time your body is warmed. Since we let the warm water flow to the sewer, the energy efficiency falls to well below 10%.

Without having to sacrifice any comfort at all, 60% to 70% of all the energy wasted in this process can be saved. A water-saving showerhead is the simplest solution. It provides the same level of comfort yet uses 40% less water, and thus 40% less energy.

But the real dark secret of energy loss in the shower lies in the discarding of warm water. Like in the kitchen sink, a simple heat exchanger could do the trick. A second water supply pipe within the shower's drainpipe will warm up the cold water for the shower. This kind of heat exchanger saves another 30% to 50% of energy consumption.

Ten years ago I could not easily acquire the materials needed to make these kinds of changes to my house. But in ten years a lot has changed. Nowadays one can buy a "heat recycler" for a shower as a product, although you may not find one in every bathroom shop. If this heat recycler is added to a solar hot water system the fossil energy used for the shower is almost zero.

A radiator on wheels

Modern day people in the West spend a substantial part of their life traveling: commuting to and from work, to a client, to the supermarket, visiting family and friends, and on holiday. For short distances the average person grabs a bike or simply walks, but the majority of all travel is done by car, plane or other (public) transportation. This requires the burning of a lot of fossil fuels.

Scientists who calculate the energy efficiency of transporting a person from point A to point B using fossil fuels (originally from crude oil) often report the astonishing low number of less than 1%!

First of all, a part of the energy is lost during the extraction, transportation, and refining the oil into gasoline. When igniting this gasoline in a combustion engine, more than half the energy is dissipated into heat. Viewed in this way, our car is little more than a glorified, fast-moving radiator.

The friction of the wheels on the road and the resistance of the air at high driving speeds virtually decimate the energy efficiency quotient. And finally, while our ulti-

mate objective is to move about 100kg of human flesh, plus a little luggage, from A to B, we actually end up moving about a ton of steel as well. And moving those extra kg's of steel make us experience other issues; expensive parking issues.

The energy efficiency of a car can be improved dramatically. There are many ways to do this. An obvious example is to drive smaller and lighter cars. Or we can put an electrical motor in each wheel. With an engine in every wheel we minimize energy losses that usually occur when bringing the required movement from the engine to the wheels. Additionally, the energy losses with an electrical motor are about a third of those in a combustion engine. And the electric motor uses no energy if the car is not moving, like at the traffic light or in a traffic jam. The required electricity comes from batteries, which of course are charged with solar power or from a fuel cell running on sustainably produced hydrogen or methanol. With these procedures the

Energy units

There is a considerable and widespread confusion about the units in which electricity is measured. The most well known unit is the kilowatt-hour (kWh), which is found on electricity meters and bills. Each household in the Netherlands uses on average around 3,500 kilowatt-hours per year in electricity, for all household appliances. With one kilowatt-hour you can use your vacuum cleaner for about an hour or run your fridge for a day.

Energy specialists commonly measure energy use in 'joule' (J), which is much smaller than the kilowatt-hour. To be precise, a joule is 3.6 million times smaller than a kiloWatt-hour. To avoid the use of too many zero's, each thousand is indicated by the prefix 'kilo' (k), each million with 'mega' (M) and each billion with 'giga' (G). So 1 kWh = 3.6 MJ.

Less common in the energy industry, but perhaps more familiar to most, is the calorie (cal). That unit is found on the label of most groceries – whether milk, bread or pasta sauce - right next to 'joules'. 1 kcal = 4.2 kJ.

The units mentioned above should not be confused with the units used to describe the capacity to provide or use energy. This capacity is expressed in watt (W). One (1) Watt is the power to provide or use 1 joule of energy for one second. So capacity is expressed in the quantity of energy used per unit of time. For ease of understanding: a small LED light uses around 1 watt, an ordinary Incandescent light bulb uses 40 to 60 watts, a human being (which acts like a "radiator") uses around 100 watts, and a large power station produces around 1000 billion watts (1,000 megawatt) of power.

trip from A to B can be made using a lot less energy, with the added bonus that the trip will be cleaner and quieter.

The doorbell

The old-fashioned and decorative doorknocker has been replaced in many households by a doorbell. A doorbell is very useful, but from an energy perspective it is essentially an ogre.

When the doorbell rings, the bell sends an electrical signal to the little appliance that produces the ring, ding-dong, woof-woof, or whatever sound may be uttered to draw our attention. This is very practical, but the fate of nearly all doorbells is that 99.99% of the time they remain untouched. For those rare, short moments of contact with a human finger it is, however, necessary to have the voltage at 100%. That voltage is a direct current (10 to 36 volt) that is converted by a transformer from the 110 or 230 volt alternating current received from the power grid. And that is where the problem lies.

A transformer continuously suffers from a small power 'leakage', typically between 5 and 10 watts, which is the amount of a charger of a mobile phone. That isn't much, but because the bell is "switched on" for 24 hours a day and all 365 days of the year, the transformer uses around 50 kilowatt-hours. That is around 1% to 2% of a households annual electricity use!

To neglect this loss would be close to stupidity, because imagine what happens to the numbers when one adds all the doorbells in the European Union. An estimated 200 million doorbells in the EU use around 10 billion kilowatt-hours each year. That is the yearly output of two fully functioning coal-fired power plants of 600 megawatt each!

To save ourselves from two power plants we really do not have to return to the era of the doorknocker. The only thing that is required is to switch off the transformer and find a way to switch it on when someone rings the bell. There is a beautiful and simple solution. As soon as the button for the doorbell is pressed, a miniscule amount of power sends a signal to the transformer to switch it on, after which it can sound the ding-dong, ring or woof.

This small source of energy is of course of a renewable nature. With the knowledge we have today it is a piece-of-cake to integrate a solar cell into a doorbell, as well as a small capacitor to store a small quantity of power. A piezoelectric doorbell button is even simpler. With a little pressure, piezoelectric materials can generate enough electricity to activate the transformer.

Here too, the energy efficiency is not determined by the relevant activity ('ringing the doorbell') but by losses in the transformer. Unless you have a remarkably large group of friends or a group of very annoying children for neighbours, a doorbell is used at most for one hour per year and the transformer is left on for a stunning 8759 hours. Not taking into account the losses in energy conversions, the efficiency of the doorbell device is a shocking 0.01%.

Take a look at how we can solve the problem I described above. We have replaced the primary energy source with solar or piezoelectric energy. By looking intelligently at the functions of a doorbell we have not made the use of 50kWh sustainable, but essentially removed the need to use this energy at all.

Energy from nine to five

The office buildings where we spend a considerable portion of our lives typically have a comfortable climate and pleasant levels of light. There are computers and printers, photocopiers and telephones. We can have lunch in company restaurants and we are supplied with cups of hot coffee and tea at the press of a button.

Most people working in office buildings don't have the vaguest idea of the amount of energy used in these buildings. In an average European office you probably use as much energy as you do at home! At home the heating, air conditioning, lights, and appliances consume around 40 billion joule (or 40 gigajoule) per person per year. In your modernized work place you use an equal amount.

There is a lot of room for technological improvement in offices. But this is not something I want to focus on. Let's look at the entire system up-close. Regardless of how hard we work, the office is only used for a third of the week. Using a simple calculation, if the office is in use from 7am to 7pm, five days a week, then we ar-

Saving ourselves power stations: routers

The examples that have been described might make you think: "these are all old appliances, surely nowadays we are developing much more efficient devices!" Nothing could be less true. In the past few years we have introduced new technology into every aspect of our daily lives. The computer, laptop, and smart phone: all interconnected via the Internet. We communicate, hold meetings, get the latest news, all on the Internet. For large parts of the day we are online and email, use twitter, are active on social media like Facebook, LinkedIn, Flickr, and at the same time stream music or videos. To provide easy access to the web, a router brings the Internet into our homes, and connects all our devices to it with WiFi or a cable.

A router uses 10 to 15 watts, 24 hours a day. However most of us are not online all that time. We sleep, go to school, and go to work. During all of those times the router is still switched on stand-by, waiting for us to come online. It is unpleasant to realise, but these modern devices in stand-by mode still devour energy: 8 to 10 watts an hour. This is absurd considering it is technically possible and easy to build a router that uses less than 1 watt of energy in stand-by mode. Let's do another calculation. In the US in 2010 there were around 82 million broadband Internet users and in the European Union there were 133 million. Each of these users has a router that could probably be switched to stand-by for 75% of the time. Therefore each router wastes around 50 kilowatt-hours a year, and the total 200 million routers needlessly use over 10 billion kilowatt-hours a year. That is the capacity of two coal-fired power plants of 600 megawatts each.

rive at a total of sixty hours. Yet in most offices the central heating, air conditioning, and often even computers are functioning for 24 hours. If that happens five days a week, that means 120 hours non-stop.

If we could switch the entire office on stand-by with a simple electronic timing device we could cut the energy consumption almost in half. If the energy consumption – and thus waste – were more visible, then everyone might become an 'owner' of the problem and the squandering could become more visible.

I know bizarre examples of squandering. One office building was heated by a municipal district heating system on a bright summer's day, while at the same time an air conditioning system was running to cool down the very same rooms. The climate inside was pleasant but the energy use was exorbitantly high. The "energy

manager" of the company had been fired previously in order to cut costs, which now proved to be a rather counterproductive decision.

Energy savings in office buildings have enormous potential once scaled to European size. My estimate is that in the entire EU over 100 million people work in offices and each person wastes 20 gigajoule of energy per year. If timer switches were used to turn everything "off" at night and on weekends this could add up to an energy savings of 2 billion gigajoule, which equals the total energy used by a country the size of the Czech Republic.

Hot controls

A few years ago I was invited to take a look in the control room of one of our typical Dutch locks. A control room like that is filled with electronic equipment for opening and closing of the lock's doors. The lockkeeper told me that summer or winter, the room was always too hot despite the climate control system and excellent insulation.

The excess heat was being generated by the control electronics themselves, and for a good reason. Since the sensitive electronics could not operate at an excessively low ambient temperature, the equipment had built-in heating devices, which were turned on day and night. The construction workers who built the central heating for the control room had not been aware of this.

The heat of the two systems had to be compensated for with an air conditioning system. This took away the discomfort but not the cause of the overly warm temperatures. Switching off one of the two heating systems would have achieved the same results, as analysis later showed.

The control room was also interesting for another reason. The equipment contained hundreds of LED lights as indicators that the machines were functioning. Only a few times a day someone might glance at these indicators. Even though LED lights are very efficient, using only 1 or 2 watts per light, the combination of 200 LED lights increased the power use of the control room by 2,000 kilowatt-hours per year. With a simple switch at the door, all these LED lights could be turned off, at least when the lockkeeper was off work. And control rooms like this exist in many industries and other facilities.

A wasteful system

Like all these examples have revealed, we collectively waste 98% or more of all the energy we extract from the ground before finally obtaining the service the energy was supposed to provide. That is not just because our technologies are inefficient, but more importantly because we do not realise how much of it we waste. Perhaps we don't care. Few people realise that doorbells, offices, and showers needlessly waste large quantities of energy.

Lighting a room, watching TV, or living in a comfortable house – all of this can be done with more energy efficiency. The change starts with the design of systems that do not throw away energy. A doorbell that only uses electricity when someone presses it, offices that remain unlit except when someone is there, or reading a book with the daylight from the outside. Let us start with these simple steps, and then later we can find the most efficient equipment and the most sustainable energy.

We must look at the energy system from the end and look at the function we expect energy to perform. This will help us conceive of many new solutions to improve our way of dealing with energy.

One hour

Being raised on a farm means growing up with the elements: the sun, wind, and water. It provides the opportunity to see plants grow because of that sun and that water. You experience the force of the wind that can uproot trees. You see the grass turn yellow after a sweltering summer's day.

Only very few people are conscious that the energy sources surrounding us could provide all the energy we need. Easily. Every year each square meter of earth receives on average 240 joule per second (watt) of solar energy. In comparison, if we calculate our world energy use in the same way, our daily primary energy use is around 0.03 watt per square meter. That means we receive from the sun almost 10,000 times more energy than we need for heating, cooling, electrical devices, fuels for transportation, and other consumer products.

If the 98% inefficiency of our energy system is factored in, the contrast becomes even greater. Were we to design a system without energy loss, the energy provided by the sun in less than one minute would be enough to meet our power needs for an entire year.

From these facts I can only draw one conclusion: there is no energy crisis. The sun is an immeasurable source of energy. If we need energy, we have to learn to look up, instead of down to the fossil fuels under our feet.

The sun is not only the source of all life, but also the source of nearly all other forms of energy. Wind energy originates because the sun warms air and makes it rise, which causes pressure differences to arise. These, together with the rotation of the earth, create wind. Hydraulic power stations run on falling water. And some of this water was first vaporized by the sun from the oceans. Waves in their turn are caused by wind. Wood and other biomass only grow if the sun shines enough. Plants are converters of solar energy into biomass, which later can be burned or metabolized. Fossil energy is compressed biomass, millions of years old, and thus even it is concentrated solar energy.

There are two exceptions: tidal energy is created by the gravitational force between earth, moon and sun. Geothermic energy arises from the deeper layers of the earth itself.

Energy flows from the sun

The strength of the irradiation of the sun is highly variable, dependent on the time of day, location, and local circumstances. Annually, the equator receives four times the amount of sun as the Polar Regions do. However, the majority of the world's population inhabits latitudes lower than the Polar Regions. It is certain that in each continent much more solar energy is available than is being used.

Here are a few clarifying examples. The United States receive on average 5,000 times more energy from the sun than national electricity consumption. Solar cells that have a conversion efficiency of 10% (currently much higher percentages are being achieved) could supply the entire US with electricity if 0.2% (1/500th) of the land surface was covered with solar cells. India, with its 1.1 billion inhabitants, needs a maximum of 2% of its land surface covered to meet its entire energy needs. And a mere 0.25% of the Sahara desert is enough to provide the entire European Union with electricity.

It is interesting to note that although currently most solar cell systems do indeed require surface where the solar cells are placed, this might soon change. Developments are under way to create systems that allow the solar cells to be integrated into roofs, windows and other surfaces. And then no separate locations will have to be used for the solar cells.

Wind

Almost 2,000 years ago humanity discovered how to use the force of the wind. In the first few centuries AD in ancient Persia, one could already find windmills. In Tibet and China the use of windmills became common around the year 1,000. Half a millennium later, windmills contributed to the prosperity of the Netherlands in the 17th century. At the peak of the Dutch Golden Age, no less than 9,000 windmills supported the flourishing economy. Rough estimates tell us that these 9,000 windmills generated a collective power of around 250 megawatt, which equals one large offshore wind farm with 80 turbines.

Channeled wind

Due to some unusual geographical features, California contains some very favorable areas for wind energy. Because of the cold Gulf Stream in front of the coast and the inland desert, a continuous sea breeze blows over the land. The hot air in the desert rises and moves toward the sea. This causes a localized low-pressure area that draws in more of the cool air from the sea. From all of this, strong winds are created and then they are channeled by the way mountains and valleys are situated. Normally wind blowing over the sea is stronger than the wind on land, but in certain parts of California the opposite is true. This situation is ideal for wind turbines, and many such turbines have been built there. These turbines cannot be too high in the air, however, since the strong flow of air is only created in the lowest hundred meters of the atmosphere.

The energy content of the wind is almost unimaginable. Someone once calculated that a hurricane at its peak could deliver up to 30 terawatt of energy, which is equal to about 50,000 large coal power plants.

An estimated 2% of all the solar energy that reaches the outside of our atmosphere is transformed into wind. We are lucky that not all of that wind is blowing in the lower layers of our air! Nonetheless the potential for wind energy is many times greater than worldwide energy requirements.

Wind at sea is a different case. The surface of the earth consists for 70% of water, which is where the wind blows more than at land. The potential of this is enormous. Let us look at the Caspian Sea. This sea is surrounded by Azerbaijan, Iran, Kazakhstan, Russia, and Turkmenistan. As the largest inland body of water in the world, it has a surface area of 371,000 km^2.

A conservative estimate is that one square kilometer in that area could yield 30 million kilowatt-hours of electricity per year. So if 10% of the Caspian Sea was used to hold windturbines, the entire electricity needs of all the surrounding countries could be met! Something similar applies to the Northsea. If windturbines were on 6% of the Northsea's surface this would use around 30,000 square kilometers and would provide the energy consumption of all the neighbouring countries: the UK, Norway, Denmark, Germany, The Netherlands, Belgium, and France!

Waves

Waves are a great source of energy in themselves. Waves are generated by wind passing over the surface of the sea. As long as the waves move slower than the wind, the waves are 'fed' with energy from the wind. A modest wave front at the Atlantic coast generates circa 50 kilowatt per kilometre of coastline. The French Atlantic ocean coastline is about 3,000 km. This could provide so much energy in wave power that it would nearly equal the capacity of all the nuclear power plants in Europe.

Waves do not only move up and down. Look really carefully at a cork bobbing up and down on the surface while waves pass below it. You can see the cork making small vertical circles: ahead, down, back and up again. Below the surface, water particles also move in circles. The deeper you get, the smaller these movements become.

There are numerous technologies under development to harvest wave energy. One of these developments is the ingenuous 'Wave Rotor'. This device is designed to 'catch' the circular movements in the water column and convert them into a propulsive force acting on tilted vertical and horizontal rotor blades.

In the waves typically occurring in the Atlantic Ocean, one 10 meter diameter Wave Rotor could yield 500 kW of electrical power. If we move only one more technological step, the Wave Rotor could revolve underwater around the mast of an offshore wind turbine and we would have two renewable sources at one location.

Biomass

Though it is perhaps somewhat less evident than wind and waves, biomass is also a form of solar energy. Plants grow thanks to sunlight. The energy in the particles of sunlight (photons) activates the chemical processes in plant cells. Then minerals, water, and carbon dioxide from the air are transformed into leaves, flowers, stalks, fruits, and roots. In short, they are transformed into biomass.

Besides this kind of biomass from plants on the land, there is also marine biomass (algae and phytoplankton), animal biomass (manure, grease and slaughterhouse leftovers) and industrial biomass (waste products from paper and cardboard factories, the dairy industry, beer breweries, food processing, and so on).

The term biomass contains within it a wide variety of materials that have one key feature in common: They do not add extra CO_2 to the atmosphere. Biomass is carbon neutral. That is a somewhat complicated term to express that during its lifetime a plant uses roughly the same amount of carbon dioxide for its growth as will be emitted into the air when the plant rots in nature or is burned. This is vitally different from fossil fuels. Oil, gas, and coal were once biomass, too. As we burn those age-old fossil fuels, we are suddenly transferring that fossil CO_2 that has been locked

Ocean Deserts for eel farming

The Sargasso Sea is a 4 million square kilometre area within the Atlantic Ocean known as the breeding grounds for North-American and European eels. Specific ocean currents keep this part of the Atlantic relatively salty and isolated from life that exists in other parts of the ocean. The area has been named 'Ocean Desert' in honour of the fact that it harbours hardly any life. There are two exceptions: young eels and gigantic amounts of the seaweed Sargassum.

The latter is a floating and fast growing mass of energy, and it is perhaps the new biofuel source of the future. Besides the Sargasso Sea there are four other of these gigantic Ocean Deserts in the Atlantic, together spanning an area of over 25 million square kilometers. Except for Sargassum, no other life survives in these "deserts". A rough estimate tells us that such seaweeds grow on less than 1% of the ocean's surface, but in that 1% there is enough biofuel created each year to fulfil the world's energy consumption 10 times over.

away for millions of years, into our atmosphere. With 'fresh' plants we are recycling CO_2 in a short life cycle of months, or at most a few years.

Biomass does not store solar energy efficiently. The process of photosynthesis and carbon dioxide uptake is not very efficient. Only 2% to 6% of the solar energy that reaches the leaves of a plant is converted into usable energy.

However, the enormous area of vegetation on the planet compensates for the low efficiency of energy transfer. Approximately 70% of the Earth's land surface is green. If we could use all the renewable vegetation growth for energy production, we could cover world energy consumption almost a hundred times over!

Of course not all vegetation can be harvested for energy. The majority is not available or hard to reach, and a large part is used as our own personal energy source: food. Biomass for energy should not compete with food crops for arable land. That is why Jatropha and elephant grass are so suitable as energy crops. These crops grow well on nutrient-poor soil where food crops would not fare well.

Algae are an even more promising energy crop. Algae are single-celled water organisms that grow in water. Many algae species produce drops of oil that can make up as much as half of their minuscule body weight. They can produce an astonishing 80,000 litres of oil per hectare. This is fifty times more than Jatropha can yield and eighty times more than rapeseed. The yield is so much higher because algae do not have roots, stems, nor flowers, and are little more than cells that convert solar energy into biomass. To get an idea of the great potential of energy production from algae, 0.2% of the ocean surface could grow enough algal oil to supply the entire world with fuel for all of its transport.

Algae can grow at sea, but also on land, in open ponds, or in closed cylinders. Like plants they live off sunlight, CO_2 and nutrients. Building gigantic seawater ponds on the Sahara no longer sounds like such a crazy idea. It would only require 0.25% of that desert to provide the entire European Union's transport fuels.
Algae are a very efficient producer of a vegetable oil that can serve as fuel for transport. They can even produce crude bio-oil which, like its fossil equivalent, can be processed in ordinary oil refineries.

Hydropower

Solar energy and gravity are the main "engines" behind the world's water cycle. It works somewhat like this: the sun vaporizes water at the surface of oceans. In the higher layers of air this water vapour condenses into clouds and falls back onto the surface as snow, hail or rain. The water flows back to the sea through rivers and lakes, completing the cycle. The amount of available energy in moving water is determined by the quantity of water and the height it falls from.

Hydropower has been used for thousands of years. In ancient India, water wheels and watermills were built. In Imperial Rome, water powered mills produced flour from grain and were used for sawing timber and stone. About 400 years ago the paper industry was able to grow by using watermills. At places where there was clean and moving water the paper industry could prosper. Around 1740 in the Veluwe (a small area in the Netherlands) there were 168 water mills used for paper production.

Hydropower and biomass are the oldest forms of renewable energy used on a significant scale. Large dams have been built in rivers to create huge reservoirs of water behind the dam. The water can be allowed to flow down through turbines in which they generate electricity. The capacity created by hydropower is currently around 800,000 megawatt, or a little less then 20% of the total world electricity production.

Still, it is estimated that two thirds of the world's economically feasible potential is yet to be exploited, mainly in Africa, Asia and South America. There are even many water reservoirs around the world – like in Algeria - that are only used for drinking water or water management, despite being perfectly capable of creating hydropower.

Large dams are still being designed, but these dams can have social and environmental disadvantages. The world's largest is the Three Gorges Dam in China on the third longest river in the world, the Yangtze River. This dam is 2,335 meters long, 185 meters high, and can produce 18,000 MW in electricity. To make this dam it was necessary to flood land up to 600 km upstream of the dam. To accommodate this change 1.2 million people had to move. Small hydropower stations do not usually cause the same social and geopolitical issues.

Hydropower is an excellent tool to generate extra capacity when there are peaks in the demand for electricity. The supply of hydropower energy can be switched on and off quickly so it is a very flexible, cheap source of energy.

At the moment most of the hydropower capacity consists of large projects with gigantic reservoirs. But, there are also many smaller projects that are simply placed in a river stream, even without a dam. In China alone 42,000 of these smaller hydropower stations are currently functioning, each with a capacity of less than 50 megawatt. Together they make up about a third of the hydropower production in

China. And there are other interesting applications to be found for these minimally invasive power stations.

Osmosis

Energy potential also exists in water. Mixing freshwater with saltwater can be a large source of energy. This 'osmotic energy' appeals to the physicist in me. Freshwater and saltwater have different physical and chemical properties, and they interact differently with their environment.

To explain how this works, let's take a quick look in the kitchen. Before frying thin slices of zucchini, your cookbook will advise you to put some salt on them and leave them for an hour. The salt will extract water from the slices in an attempt to equalize the salt concentration inside and outside the zucchini. The tendency of nature to eradicate differences in salt concentrations is called 'osmotic pressure'.

The same principle can be seen at work in two reservoirs of water with different salt concentrations. In between those two reservoirs we place a membrane with holes that are so tiny that they allow water molecules to pass through, but not the larger salt molecules. The osmotic pressure wants to equalize the salt concentrations on either side of the membrane and this can only be done in one way: by allowing water molecules to move to the more salty reservoir where they will dilute the solution. So on that side of the membrane the water level is raised until the water pressure exactly matches the osmotic pressure. Then the two reservoirs are in balance.

The energy that is generated by osmotic pressure between saltwater and freshwater is surprisingly great. The pressure difference between seawater and water from a river is equivalent to the pressure exerted by a column of water about 270 meters high! This pressure difference could be used on a large scale in the estuaries of the world's large rivers.

Again, the sun is the driving force behind this form of energy. It is the sun that

Reversed osmosis for seawater desalination

Osmotic energy is a remarkable, young technology, but the reverse process has been in use for years: to desalinate seawater and to purify wastewater. In this reverse process, freshwater is separated from saltwater by adding energy. The use of reversed osmosis for desalination is a multi-billion dollar business and is still growing rapidly. Thousands of these reversed osmosis units are already operating. The technology is available, and to apply it to osmotic power generation would not be too complicated.

causes water to evaporate so that a salty sea remains. And via the water cycle the water that has evaporated flows back into the sea.

Theoretically, each cubic meter of water flowing out to sea could produce 0.2 kWh of electricity. Counting all the rivers, the global discharge of fresh water to the seas is about 44,500 km3, or 10,000 billion kWh of electricity per year which is approximately 60% of global electricity demand.

Locally, yields may even exceed demand, like the estuary at the Amazon River, which alone could generate more electricity than Brazil needs. Bangladesh could be an exporter of electricity if the Brahmaputra and Ganges rivers were used in this way. It would be even more practical to build an osmosis pump that could pump dry the lowlands of Bangladesh when they are flooded. In situations like these the water will have to be pumped up a few meters, while the osmotic pressure gradient is more than 200 meters. Bangladesh could remain dry and at the same time be able to produce electricity.

Tidal energy

Besides the energy from the sun, we have gravity as a source of renewable energy. The gravity of the Earth, Sun, and Moon provides us with large amounts of energy in the form of tidal movements. The ocean's tides are driven by the Moon's cycle around the Earth. Sometimes this is reinforced or weakened by the gravity of the Sun and by the wind.

Four times a day enormous amounts of water in the oceans and seas are displaced, two times back and two times forth. In some locations the tide pushes water through narrow passages and creates strong currents. Every day ships in harbours are lifted several meters up, and later returned to the mud.

In this context, the Bay of Fundy deserves special mention. Lying between Nova Scotia and the Canadian province of New Brunswick, this bay is renowned for the huge differences between high and low tide: up to 20 meters which is as high as a six-storey building! Twice a day this natural reservoir fills up with about 300 cubic kilometres of water and then empties again. By building a barrage that only harnesses the top 4 meters of this tide twice daily, for about 2 hours at a time, a fifth of Canada's total electricity demand could be met.

Earth core heat

The inner thermal energy of the Earth is a third major inexhaustible energy source at our disposal, along with the Sun and gravity. For five billion years remnants of the solar system's birth have been kept safely insulated within the Earth's mantle and replenished by radioactive decay they will persist for at least another seven billion years. Tapping some of this heat will have no negative effects on the planet. In the outer 10 kilometres of the Earth's crust beneath the United States lies some 30,000 times the total annual global energy consumption.

Geothermal heat is most tangible in volcanic regions, where heat from the Earth's depths comes to the surface. One of the most remarkable forms of geothermal heat is a geyser. At some 100 locations around the world, the combination of water, heat, and geological structures create this breathtaking spectacle; like in Chile, Yellowstone National Park , and Iceland.

In countries where volcanic activity is high, like Italy and Iceland, the extraction of geothermal energy is already a proven technology. In these areas, temperatures underground are high enough for steam to be generated simply by injecting water into the fractured 'hot rock' and extracting the resulting steam via a pipeline toward a turbine. From less deep and less warm layers of the earth's crust it's usually the heat itself that is used directly for heating.

Geothermal energy is everywhere. I was struck by the promising and profitable conditions in Poland. The underground temperature there rises by 1° to 4 °C for every 100 meters of depth. Between one and three kilometres into the earth's crust the temperature is between 30° and 130°C. This represents a monumental reservoir of energy: over 20,000 times Poland's total annual energy consumption. But in Poland there are only a few geothermic power stations. The most recently built plant was in Mszczonów in 1999.

Everywhere

In our prosperous communities with easy access to commercial energy we have largely forgotten that it was solar energy that powered the development of humankind. Before discovering fire, we lived on the solar energy that was stored in our food and warmed our caves. The discovery of fire led us to use wood for cooking as well as for making tools. But then again, wood is merely solidified sunlight.

It was only late in the Dark Ages, at the end of the 15th century, that Western civilization found out how to use fossil fuels. As the agricultural societies of the west became urbanized, their populations needed more concentrated fuels than wood or charcoal. Peat – partially decayed vegetation that can be seen as an early stage of coal – provided the first cities in Western Europe with the fuel they needed to survive through winter.

That was the start of the era in which we began to use up resources faster than they could be replenished. Large areas of peat land in England, Ireland, Finland, Holland, and Germany were emptied rapidly.

The era of peat was also the beginning of our addiction to fuels from the Earth's crust. 'Addiction' sounds negative, but fossil fuels have also brought us much prosperity. Roads and canals were built to bring peat to cities. Gradually, from the eighteenth century on, open fireplaces in houses were replaced by stoves and ovens. This was a great innovation. In 1742 Benjamin Franklin himself invented the metal freestanding stove with air circulation, to allow for a better combustion of peat and later coal.

Fuels that contained a higher concentration of energy continued to be one of the main features in the development of Western civilization's energy demand. The Industrial Revolution was only made possible by the concentrated fuel contained in coal. From then on, technological development and the mining of fossil fuels went hand in hand.

In some countries, like China, Spain, and the Netherlands, wind energy helped pave the way for the Industrial Revolution. But windmills soon became obsolete. In other countries hydropower played the same role, and this sustainable source of energy survived the Industrial Revolution. At the same time our move toward fossil fuels was accompanied by a further centralization of our energy supply system.

Because of these changes, the development of alternative sources of energy was greatly overlooked. The direction of the change was clear: away from the sustainable sources in our own environment that we had once used. Those sources were simply no longer needed. That situation is now rapidly changing. And with current knowledge and technology, which have started to expand geometrically, we can use sustainable sources in a new, efficient way.

This chapter presents some surprising examples of large sustainable energy sources in places where we had not expected them to exist. Renewable energy is literally everywhere, ready to be used. The golden rule is: look around you, energy is everywhere.

Waste

Although the idea may sound strange, manure (both animal and human) is another clean source of energy. Industrialized countries are so detached from the agricultural way of life, that we generally deny the energy potential of manure. In many rural areas across the world the value of manure is still recognized. While countries have generally enacted legislation to eliminate the health hazards that are contained within faeces, these usually do not take the energy potential into account.

Dried manure is a perfectly fine solid fuel, but by using a process called 'anaerobic digestion' another whole range of possibilities is opened up. Under specific atmospheric circumstances microorganisms can break down the waste along with any hazardous organic material in it. Odours, pathogens, and viruses are broken down while large quantities of biogas are produced, along with an excellent fertilizer. The resulting biogas is very similar to natural gas, and can be used to generate either electricity or heat. A beautiful technology, anaerobic digestion is nothing more than an engineered, optimised version of a naturally occurring process.

Livestock produce manure day in and day out the world over. Together these animals essentially form an enormous biogas factory. A single cow produces about 26 cubic meters of dung per year, or the volume of an average bathroom. Using anaerobic digesters, one cow could supply about 10% of the heat demand of a modern house in northern Europe. By adding so-called 'co-substrates' (e.g. readily available organic waste from the food industry) net production could be increased significantly.

Why not think a little bigger? Across the planet there are 1.5 billion cows. From their manure we could make biogas with an energy value equal to 260 billion cubic meters of natural gas. That is half the amount of natural gas used by the European Union or the total gas used by Germany, the UK, France, and the Netherlands.

Paunch manure
It takes an open mind to see that manure is a clean source of renewable energy. Nevertheless, we have other and arguably even stranger sources of renewable energy. Every day many pigs, cows, and chickens are killed and processed into the consumer-ready meat we know. Worldwide, mankind annually slaughters 50 billion chickens, 1.25 billion pigs, 275 million cows, and over 525 million sheep.

When these animals are slaughtered they have a considerable amount of food in their stomachs and intestines. This organic material is also called 'unborn manure' or 'paunch manure.' This material, after collection, produces twice as much biogas as normal manure. That is because there is more energy contained in undigested food than in faeces.

Here are some calculations of the potential of this unusual source of renewable energy. Each slaughtered pig contains about 4 kg of paunch manure, and every ton of this matter could produce 45 cubic meter of biogas. All the paunch manure of all the slaughtered pigs in the world can produce over 4,500,000 billion joule per year. This is plenty of energy to boil one egg for each and every person on the planet, every week.

The Pine Beetle
Canada, the second largest country in the world, is largely covered by ancient forests. The lumber industry has been a key element of the Canadian economy, and the government has pushed for sustainable forestry in order to allow for the long-term harvesting of their forests.

In recent years, vast tracts of these forests have become infected with the Mountain Pine Beetle, an unsightly 5mm long insect. This small beetle lays its eggs just under the bark of a pine tree. Because that is the same place where the trees nutrients flow, the tree is cut off from its lifeblood and dies within weeks.

The consequences are truly disastrous. As of 2008, 14.5 million hectares of boreal forest (4% of the country's boreal forested area) had been defoliated by this pest. The Pine Beetle is a naturally occurring organism in the Canadian woods. In the past the number was controlled by nature because the small beetle cannot survive consistent temperatures below minus 20° C. As a result of climate change the temperature in large parts of Canada has risen so much that the Pine Beetle now comfortably survives year-round.

The sustainable use of the Canadian forests is in jeopardy. These beetles have to be controlled. All of the dead wood in affected areas has to be removed, but the quantities are far greater than the lumber industry can handle. In some cases sections of forests are deliberately set on fire to combat this plague.

Could this wasted wood perhaps be used differently and more productively? The wood from these dead trees is a great source of biomass. Most of the moisture has already left the wood. If all the dead pine trees in the affected 14.5 billion hectare area had been used for electricity production, an estimated 300 billion kilowatt-hours would have been harvested. That is 60% of Canada's yearly electricity consumption. And naturally, the dead trees would have been replaced by careful replanting to support a sustainable lumber industry.

To put this number in a different perspective, the 14.5 million hectares of land could sustainably generate the same amount of energy as the oil currently produced by the Canadian tar sand industry. The crucial difference being that such forests will supply biomass forever, without increasing the country's greenhouse gas emissions.

Empty fruit bunches
Palm oil is the most popular edible oil in the world. It is used in toothpaste, margarine, and many other products. It can even be processed into diesel fuel and used in engines. In Malaysia alone there are 4.3 million hectares of palm oil plantations. Each year these plantations produce 16 million tons of palm oil, which is 40% of the world production.

Each "bunch" of palm oil fruit is about the size of a football, weighs about 25 kilograms, and contains up to 7 litres of oil. The empty fruit bunches, the majority of the fruit, are currently considered waste and thrown away. This too is renewable energy. Malaysia generates over 5 million-tons of this dry matter annually. Using this biomass as a source for power generation could easily cover 15% of Malaysia's electricity consumption and a significant portion of its industrial heat demand. In the future, with new technologies already under development, these waste products could even be transformed into clean, sustainable biofuels.

Cold water

Airconditioning with seawater
Sun drenched shorelines across the world are filled with large hotels where hordes of tourists luxuriate in the warm weather, take a refreshing swim, dive and snorkel, and find a cool place to dine and recover from the heat. These comfortable hotels are usually equipped with very large air conditioning units.

Such air conditioning systems commonly run on electricity. At larger hotels an electrically driven chiller is placed near its buildings. These chillers produce cold water that is pumped through the building in order to cool the air in rooms, halls, and other spaces.

At most locations around the equator, such as the Caribbean, temperatures hardly vary during the year, making it necessary to cool the buildings year-round. A staggering 50% of total electricity consumption in these tropical areas is used for cooling. To make things worse, this electricity is often provided by small generating units with very low efficiencies. Along with electricity, they also produce smells and noise.

Now that we have discovered that sustainable energy can be found anywhere, can we think of a solution to this problem of high energy consumption? The cheapest and easiest solution is not found in the sun and wind, which the Caribbean of course does not lack. Instead it is found in the ocean. Most hotels and other structures have been built near the beach, where sunbathing tourists can have their occasional swim in the shallow waters of the sea. A little further into the ocean there are vast amounts of cold water. Simple technology entitled 'Swac' can extract this cold from the ocean and prevent a lot of needless electricity generation.

'Sea Water Air Conditioning' (Swac) technology is little more than a pipe in the sea that transports cold seawater to the shore. Water with a temperature of about 6° C is brought to shore. In the warm waters of the Caribbean this requires going to a depth of 700 meters. A simple heat exchanger extracts the cold from the seawater. This cold is fed via a closed-loop fresh water system to bedrooms and halls. Then the used seawater is transported back to the ocean to the exact depth were the temperature is the same as the used water, which is about 12 °C.

Swac is no more than a variation of heat or cold storage in the Earth's crust, so it is not particularly revolutionary. But using the sea as a source of cold is relatively new. This simple proven technology could reduce the energy used for air conditioning by 90%. A pump is hardly used, because the 'siphon principle' does all the work. Only an initial moment of suction is needed to start the water flow. This works just like the tube in the jerrycan that I used to 'borrow' gasoline from my neighbour's motorcycle.

The cold ocean water mentioned here is also a renewable energy source. Water in the oceans is continuously refreshed and cooled by the ocean currents that originate in the poles and are spread around the Earth. Since cold water is heavier than warm water, it forms an undercurrent. All that one needs in order to profit from this natural phenomenon is a sturdy pipeline, a little knowledge of physics, and everything else comes for free. Countless applications of this are possible.

Drinking water
The use of Swac is not limited to air conditioning hotels. After cooling the hotel, the warmer but still cool water can be used to irrigate a golf course. The water of Swac cools the ground and the air above it. The water molecules in the air condense just above the grass, making traditional irrigation unnecessary. This is the same simple principle that explains the drops of water on the surface of a cold bottle of beer on a hot day.

In the tropics, a Sea Water Air Conditioning system with a 40-megawatt cooling capacity is enough to cool several large hotels and office blocks. It could also produce

800 cubic meters of drinking water daily, using the same principle of condensation as on the golf course. That is enough to meet the needs of three hotels with 500 rooms each, or to irrigate two 18 hole, 65 hectare golf courses.

Cold electricity

A third possible use of the ocean's cold water is direct electricity production. If the temperature difference between surface water and deep seawater is more than 22 °C, Ocean Thermal Energy Conversion (OTEC) can generate electricity. The potential of this application is mind-boggling. Over five times the global electricity demand could be generated with OTEC. The first candidates for this technology should be those areas that use a lot of energy, like industrial locations or coastal cities.

In the design of such a system we should be bold. If the cold seawater is used for all three functions (to generate cooling, freshwater, and electricity) the costs for the infrastructure of each function would be significantly lower. When building such a system, it doesn't matter much how wide we make the pipeline.

These examples show that a completely sustainable energy system can be designed using deep sea water. And finally, water from the deep sea contains high levels of key minerals, perfect for farming fish, lobster, or shrimp.

Solar Internet

Some two to three billion people around the world still have no access to electricity in their home. They use kerosene lamps for lighting, which is expensive, unhealthy and dangerous. TVs, radios, and fridges – if present at all – run on old car batteries, which need to be recharged at car repair shops many kilometers away. This process is expensive, cumbersome, and environmentally unfriendly. Mobile phones may have conquered the world, but in these regions their batteries cannot be recharged at home.

Governments and Non-governmental organisations tend to argue that connecting these households to an electricity grid would help increase their living standards. That is most likely the case, but electricity grids are expensive to build and can take decades to start functioning. There is a different approach. By using the potential of renewable energy sources like the sun, people could power their own houses without having to wait for years. Solar energy is free and available in abundance.

Naturally, this is not a new idea. Many individuals, Non-governmental organisations, and governments have tried to supply local villages, communities, schools and hospitals with solar panels to power lighting, televisions, and fridges. Those panels, and their accompanying batteries, were often given away for free.

Sadly, many such projects have ended in failure. Although solar panels were installed, there was no knowledge or organization to do the necessary operation and maintenance. Sometimes a simple broken cable was enough to leave the entire system unused. And then it starts again, the inhabitants of villages carrying their old batteries to the car repair shop kilometers away.

Wheeling in some solar panels is clearly not a solution. We have to take an interest in the needs of the panel's users. What do they want? What will improve their quality of living and the community's economic situation? How can all this remain affordable? In effect we have two goals: sustainable energy and local business.

We started to think about this concept a couple of years ago. We didn't want to simply provide people with solar panels or just energy, so we sought to provide them with services and products they needed or simply desired. People around the world want to move forward. Studying, phoning relatives abroad, or watching a football match can all be done fairly cheaply because of the Internet, which is now available even in the most remote areas. The problem is the lack of electricity.

An Internet center that runs on solar energy is the solution.

The solar powered Internet cafe we have developed is conceptually robust and very energy efficient. The café has energy efficient equipment, powered by solar panels, and accompanied by a storage system. A local entrepreneur is running this center on a for-profit-basis, aiming to make a living out of it. He provides services such as distant learning, telecommunication, Internet access, government information (for example on AIDS and malaria prevention), movies, and football matches. So far selling tickets for the football matches has proven the most profitable.

The local entrepreneur runs the cafe as a franchise. An international company provides the equipment, the renewable energy system, training, maintenance, services, and marketing support. At the moment, several of these solar powered Internet centers under the name of NICE are successfully up and running in Gambia (West Africa) and expansion to Zambia and Tanzania is imminent.

Energy from movement

In 1992 the shoe company L.A. Gear started marketing shoes they call L.A. Lights. While the name might not ring a bell, the image of grown men and women walking around with shoes that light up with each step surely comes to mind. While these shoes went out of vogue as quickly as they entered it, some interesting technology was involved: piezoelectricity. These shoes did not have batteries, but a piezoelectric element to produce the power needed for the lights.

Piezoelectricity refers to the ability of some materials, notably crystals such as quartz and sugarcane, to generate a slight amount of electricity when they are compressed. In the case of shoes, such pressure obviously comes from each foot as it touches the ground. Although the electrical energy released by the crystal is very small, it is sufficient to power the efficient LED lights that are built into the shoes for a second or so.

Piezoelectric elements are currently used in the spark igniters contained in ovens, sensors and loudspeakers. The energy intensity of current piezoelectric crystals is fairly modest. But new crystals with higher power output will soon be used to convert kinetic energy into electricity. Imagine what you can do with those! A dancing crowd could produce power for the DJ's equipment as people move and jump across the piezoelectric dance floor. And what about a piezoelectric element under each key of your computer keyboard or mobile phone? If you work hard enough on your computer or send a lot of text messages you could recharge the battery of your computer or phone. Or you can make piezoelectric carpets. You could place these carpets at places were a lot of people pass by every day in order to generate electricity. In Japan there is already such an experiment with turnstiles in a subway station.

The ultimate use of this idea would be a wind generator that does not move. You could place piezoelectric elements in the outside walls of buildings and the turbulent wind would compress piezoelectric elements and generate electricity.

Everyone

Producing energy costs money. You need a system to convert primary energy into a consumable form. From that perspective there is not such a great difference between fossil and renewable sources. At every scale something is needed to convert energy, as well as cables or pipes to bring it to the place where it is needed.

Building a wind farm is relatively more expensive than building a coal-fired power plant. However, such a fossil-fuelled plant needs a constant supply of coal, gas, or oil. These have to be extracted from one of a limited number of locations and they must be bought. So these fuels cost money.

With sun and wind the situation is different. These are available freely, everywhere, and (almost) always. Dependent on where you live, water, plants or other biomass may also be freely or cheaply available. Certain regions may have more sustainable resources than others, but each region has plenty for itself.

There are two negative characteristics of renewable energy that deserve some attention. First, the harvesting of sustainable energy needs a lot of space. Sun, wind, and plants deliver a dispersed form of energy in comparison with fossil fuels. Second, the renewable energy sources are often not constant. The sun only shines during the day and with varying strength, the wind occasionally does not blow at all, and plants grow rather slowly if it is cold. In other words, sources of sustainable energy are intermittent.

On the basis of this knowledge, how can we design a sustainable energy supply system for everyone? Such a system will be markedly different than our current system; that much is clear. While fossil fuels might be expensive and in limited supply, they are also very concentrated and controllable.

Therefore the design of a sustainable system does not start with the question: where do we find those sources? It starts with the question: what is it that we want to have? A comfortable house? Then we will design such a house on the basis of demands for comfort, but in the most energy efficient way. Only then do we will look for renewable energy sources in our house and our environment, including possibilities for energy storage. If this turns out not to be enough, we will move away from the home and look into using renewable sources further away, which perhaps we share with others.

This is how you design and develop a sustainable system: from demand to source, from local to central, and from consumer to energy company. Not the other way around.

What does a sustainable energy system look like? There is no simple answer to this question. There are many ways to build such a system and they are primarily dependent on local circumstances.

We start by categorizing the energy services that fit the reality. The demand for energy can be divided in four areas: cooling and heating of buildings, transportation, electricity, and industrial processes and resources.

Let 's dive a little deeper into each of these.

Our house

Heating, cooling, and ventilation (or more succinctly: climate control) are responsible for the bulk of the energy consumption in buildings.

In its very essence there is something quite strange about the way we use energy in our buildings. Simply put, in the winter we want to drive out cold, while in the summer we want to get rid of heat. In this way we also dump warmth during the day that we would have been able to use in the evening.

At the same time that a building uses a large quantity of net energy, the sun sends more energy to that building than it could possibly use. It would be more logical if a building was not a net energy consumer, but an energy producer.

How can we move to a situation like that? By using the life cycles of heat and cold. Actually there are two such cycles: the short cycle of day and night, and the longer cycle of the four seasons. The missing link is energy storage.

Let's start at the user and his/her immediate environment. Each person is a radiator

Walking energy

We human beings are walking energy sources. We consume energy that we call 'food.' We use this energy to maintain a stable body temperature and to allow us to move. At the end of the food's journey through our body we are left with faeces – poo. Every day our average consumption is 2,500 kcal (one calorie is 4.2 joule), which means that in a year we consume 3.7 billion joule (3.7 gigajoule) per person. We annually produce 3 billion joule of heat as a "radiator" of around 100 watts. Of the remaining 0.7 gigajoule we use 0.3 billion joule to keep our body temperature stable and to move. The remaining 0.4 GJ per year becomes poo.

of 100 watts. Computers, TVs, fridges, and other appliances together generate several hundreds of watts in heat. Added to this is the energy in our movement, from every step we take to every button we press. And the biomass that we leave in the kitchen or in the toilet is a source of energy, too.

It is a shame to waste all that heat radiating from our bodies and our appliances. With good insulation, a controlled ventilation system, and a heat exchange device we can use most of it. In the summer we want to achieve the opposite, but we can do so with the same solution: good insulation and a controlled ventilation system. Next, we use the ground under the building to store heat in the summer and cold in the winter. By exchanging heat and cold at the right times, the natural mismatch between heating-cooling supply and demand can be solved. By doing this we can reduce the remaining energy demand to only ten per cent of what it is now.

This phase shift in daily and seasonal cycles can be achieved directly, by venting the incoming and outgoing air through the ground under the building. We can also do this indirectly with the help of a heat pump, which functions as an amplifier. A heat pump can boost modest ground temperatures to levels suitable for heating, or it can lower modest ground temperatures to levels suitable for cooling.

Only after we have fully benefited from these opportunities - and no sooner – can we start to use the sustainable energy that may be found in the vicinity of the building. The list of possibilities is long, like we have shown in previous chapters. Even if there is an energy problem in buildings, that problem is clearly not a supply problem. The primary challenge is to connect the correct form of renewable energy to the system.

The 'passive house' proves that it is no longer necessary to use energy for heating and cooling our houses, offices, and other buildings. When design alone cannot fully achieve this, there is always the possibility of using a storage system. This is based on the simple idea that storage in summer and winter is sufficient to keep a house comfortable throughout the seasons. The example of the 'Closed

Passive house
The ultimate application of the above concepts in houses is the 'Passiv Haus'. It is no coincidence that the name of this concept is German, because the invention first became available in German niche markets. A 'passive house' remains at a comfortable temperature without conventional heating systems. The leading role is played by high-quality insulation and controlled airflows within the home. By making efficient use of internal sources, the energy consumption is reduced to practically zero.

Greenhouse' in the first chapter makes us aware of the fact that buildings can also be net-producers of energy. That surplus of energy could be used to supply the less efficiently designed houses, schools, offices, theatres, swimming pools, and so on with the heat and cold they need.

Our car

Global electricity generating capacity in power plants currently stands at about 4,000 giga (=billion) watts. Over the next five years this figure is expected to grow to about 5,000 gigawatts. The question is, will we invest in extra capacity or in other solutions?

First we must realize that all the power plants together do not represent the largest available capacity in the world. From an energy perspective, the fleet of cars gracing our planet is many times larger. Conservatively calculated, the world's 'capacity on wheels' is a considerable 50,000 gigawatt in engine power (1 billion cars times 50 kilowatts on average). With an efficiency of only 30% one still arrives at a remarkable 15,000 gigawatts.

Of course the comparison is a false one, because cars do not produce electricity. The number is, however, a good indication that we may have left very large possibilities unexplored. 65 million new cars come off production lines annually, each with an average power of 80 kilowatts. That means that every year we are deploying onto our roads a new power system with a gross capacity of 5,000 gigawatt – more than all the world's power plants put together.

Given the fact that the average car is out on the road for only 7% of its lifetime, it is wise to think about how cars could be used in the other 93%, like generating power. The only thing we need to do is design car engines that continuously produce electricity and connect them to the grid at home or at work.

This basic idea is nice, but it of course needs some polishing. Today's cars are not particularly energy-efficient. These "mobile heaters" still run on fossil fuels. But with an added fuel cell the cars of the future could have a very efficient small-scale generating plant under the hood. Even in partial load, a fuel cell yields power with an efficiency of over 50% and runs on hydrogen or methanol.

Such a car would be an extended version of the hybrid vehicles that are already proving so successful. While the car is driving, the fuel cell provides the power for the electric motor; while parked, it feeds power to the grid. And if the cars run on hydrogen or methanol derived from biomass, an entirely sustainable electricity production system has been created.

Together these cars could form a system of virtual power plants and at the same time a gigantic storage facility. This decentralized system with millions of cars has low net energy losses, is very efficient, saves us from making large investments in power stations, and can easily compensate for the variable nature of the sun and wind.

Of course all this will require a certain amount of technological innovation, as well as electronic and economic innovations. There is a plethora of calculations that will need to be made, with the results indicating who will pay the bills and who will take the profits. But looking at it in purely technological terms this challenge is at most a modest one, if compared with inventing a technology that harnesses solar energy. And after all, we managed to figure that one out.

Nowadays we have separate energy systems for electricity and for transport, but with the electric car, batteries, and fuel cells we can couple the electricity system with the transport fuel system. With millions of cars and millions of solar panels, we could create an energy system that functions somewhat like the Internet. Everyone can be an energy producer, consumer and trader at the same time. It is apparent that this system is diametrically opposed to the present fuel and electricity systems. The change, however, could happen quickly. Like I explained in the beginning of this chapter, in the space of one year we could – if we wanted to – have an entirely new energy system up and running. That enormous collective power generator would, however, be driving on our roads.

Our house and our car coupled

Our house can consume and produce energy. Our car can consume and produce energy. A connection in the form of a sort of energy-internet sounds futuristic. And it raises the question: Is this going to have an effect on our comfortable daily lives? Let's design a new housing block based on these principles and today's technologies. Our starting point is that all the energy we need for our house and for transport can be found in and around our house. The energy system will be an all-electric system: for heating and cooling, for all energy used in and around the house, and for the car.

We begin with a house that has a comfortable climate to live in. That does not mean we have to build a PassivHaus. We can build a house with other techniques that have proven to work. It is a compact house, facing south for optimal light capture, and has good insulation. The air is ventilated and the heat is recuperated from the ventilated air. The floor provides warmth in winter and coolness in summer. A heat pump provides the necessary heat and cold, fed by the stored energy in the ground under the house. In this way a Northern-European household would need no more than 2,000 kilowatt-hour per year in electricity to create a pleasant in-house climate. Further we will need electricity for light, hot water, cooking, and appliances. Wherever it is possible we will let daylight stream inside. At night we will use LED lights. In the sink and in the shower there will be a heat exchanger. We will cook with a microwave and induction. Of course we will cook our egg with high-energy efficiency, maybe even in the microwave.

The laundry machine and dishwasher are supplied with warm water directly, if warm water will still be needed after the development of modern detergents. All appliances are highly energy-efficient. Smart control and domotics monitor the energy efficiency of the house, and on the door there is a doorbell that runs on solar power. The total energy use of appliances is no more than 3,000 kilowatt-hours per year.

Around the house we use energy in the garden and on the street. A robot, solar powered lawn mower cuts the grass, and the fountain and garden LED lights run on solar energy. In newly built areas street lanterns are superfluous. LED lights integrated into the road provide visibility and safety and are powered by solar panels also integrated into the road. Other safety and lighting systems are directly linked to solar cells, so no connections to the grid are necessary.

We hit the road with an electric car. Clean and quiet with ever-improving batteries and a fast-growing driving radius.

The car is now part of the electricity system. At night the car is placed in a 'docking station' in the garage or in front of the house, where the batteries are charged. During the day the car is in use, or the batteries of the car can deliver electricity to the house. A smart control system makes sure that supply and demand are matched, for example per block or per borough. At central parking lots fast chargers are available. Assuming that the car drives 20,000 kilometres per year, its yearly energy usage is about 3,000 kilowatt-hours.

To summarize, per year we consume 2,000 kWh for heating and cooling, 3,000 kWh for appliances, lighting, hot water and cooking, and 3,000 kWh for driving our electric car or 8,000 kWh in total.

How will we acquire this energy? Sunshine is absent at night, and weaker in the winter than in the summer. For the wind the opposite holds true. Therefore we will generate half the electricity with solar panels and the other half with small, locally placed wind turbines. In the Northern part of Europe, to produce 4,000 kilowatt-hours of solar power we need about 50 m^2 of solar modules. A large portion of that can be placed on roofs and in walls (say around 30 to 40 m^2), so we have to find

10 to 20 m^2 of space in the backyard, the parking lot, or in an 'energy wall' along the road.

The remaining 4,000 kWh is produced with small wind turbines. These do not resemble the larger, more familiar wind turbines. Normally they have a vertical axis instead of a horizontal one. On the roof they resemble a modern chimney. Their rotor diameter is between 2 and 4 meters and it can best be installed at a height of 10 meters. Dependent on its size and height, each house needs one or two of these to produce 4,000 kWh annually.

The electricity grid in the house would ideally be direct current (DC) instead of alternating current (AC). Most appliances work on direct current and are equipped with a transformer. Solar cells produce direct current, as do batteries.

At the moment we convert everything into alternating current in order to be able to connect it to the grid. If we start using solar energy to supply a computer we convert twice: from the solar cell to the grid and from the alternating current grid to the computer. If we store the electricity in a battery there would be four conversions from AC to DC or the other way around. Conversion costs us only a few percent of energy each time it takes place.

You could use the electricity produced by solar cells directly in your house or store it in your batteries. If that is not possible it will be fed to the grid of the area. A smart system would first supply your neighbour, or, if he/she does not need it, the next house. Only if the power can find no user in your residential area would it go to the central power grid. This would require smart meters, smart switches, and smart billing-procedures.

Please note, all the technology needed for our 'smart living area' is already on sale! The only innovation that is needed relates to designing and the combining several energy systems. I would dare to make the bold statement that this system would lead to a higher level of comfort than the traditional systems. The house is always comfortable, even cool in summer. There is daylight in all rooms, including the bathroom. There is no need to visit gas stations because the car is charged at home. It is quiet around the house, because cars make no noise. And the grass is mowed automatically.

Our electricity

So we can develop a new housing area that produces as much electricity as it consumes, including the electricity for its inhabitant's cars. However, the houses in this area are large, stand alone houses, with some space around them. This is not a city

The green car park
Connecting the house and personal forms of transport offers us unusual new possibilities. What about the function of a company, university or residential area car park as a local power plant?

When we power our car with electricity we use batteries. In the future we will also use fuel cells. These can turn the energy in fuel into electricity with very little loss. Each car would then be a very efficient small power plant.
So it would not be such a mad idea to supply biofuel to multi-storey car parks containing hundreds of fuel cell equipped cars, and in this way use them as mini power plants. Given that there are 500 cars in a parking garage, each with 80 kilowatt capacity, we will have an efficient energy generator of 40 megawatt. Fifteen of such car parks will equal the capacity of an entire coal-fired power plant. That means we wouldn't have to build that plant anymore, we will have bought it already.

area with blocks of apartments. Although new apartment buildings do not have to consume a lot of energy, they will not be able to generate all the energy they need. Somehow this deficit has to be resolved.

Moreover, many people currently live in houses or apartments. These houses are not well designed from an energy perspective. Of course these houses could be renovated, but that is not enough. So for these existing dwellings we will also need power generated elsewhere. The same applies to all existing office buildings, shops, churches, schools, and other buildings. And finally, industrial processes will also need a considerable amount of energy.

I have already described a system in which cars do not only get charged at home, but also produce power with a fuel cell. This is one possibility for the generation of electricity, although the fuel for these cells will have to be produced too – preferably in a sustainable manner. That requires extra capacity outside residential areas.

What we need is space. And where can we find more abundant space than at sea? The ocean's vastness is a perfect match for the dispersed character of sustainable energy. Oceans and seas offer an ideal possibility for the large-scale production of renewable energy. Naturally the ecological value of the sea must be maintained.

It gets even better. The sustainable sources at sea are more intensive than on land. At sea the wind is not slowed down by obstacles. Algae growing at sea have considerably higher energy yields than plants on the land. Moreover, at sea there are sources that do not exist on land, such as wave energy, tidal energy, and osmosis energy that exists on the border of fresh and salt water.

Large-scale energy production at sea is a perfect addition to the decentralized system on land that I described in this chapter. What might such a system at sea be like?

Poseidon

I have chosen the Greek god of the sea as a symbol for an energy system at sea, a system that could be built gradually. Its building blocks are the enormous sources of renewable energy. In this system, offshore wind farms generate large quantities of electricity. Wave and tidal energy are extracted on a large scale, while big 'powerhouses' at estuaries harvest osmotic energy. In the coming decennia there is also space for fossil power stations, naturally equipped to capture and store CO_2.

These different energy sources will be connected to each other and to the land via an offshore power grid, specially designed for these types of energy supplies.

Wind turbines on the sea have great potential and have few limitations. The first wind farms were built in shallow waters, but a floating wind turbine is already being developed.

Wave energy is found in the same regions as wind energy. The yield of wave energy is lower due to lower waves. But we can make savings and compensate for this by building the two systems alongside each other. It is possible that a wind turbine can even be built on the same foundation as a wave energy generator.

Biomass can also grow in the same areas. For example, you can simply place a rope between two turbines and let seaweed grow on it. Of course there are many practical issues and complications, like how does one harvest the seaweed, but these are solvable.

In the past decades 'energy from the sea' meant 'extracting oil and gas'. Clean fossil power plants are another element of Poseidon. Gas and oil plants could generate electricity at sea, and the CO_2 that is produced can be stored directly in the emptied oil or gas fields. This way we solve one of the main objections to fossil fuels: their effect on climate change.

The techniques involved in pumping CO_2 into an oilfield are widely used, mostly in order to improve the productivity of the oil source via 'enhanced oil recovery'. Sometimes CO_2 is produced just for this purpose!

Once the fossil offshore power plant has finally finished with its fossil tasks, it can be used to process algae oil or other biomass. If we continue capturing and storing CO_2, the system not only prevents CO_2 emissions, it actually reduces the net amount of greenhouse gas in the atmosphere! Biomass is carbon-neutral, so storing CO_2 that results from generating biomass power creates an effective 'negative emission' (it takes CO_2 from the air).

The infrastructure that divides up power generation and bringing this power to land lies at the heart of Poseidon. There are two key aspects here. First there is the connection between the networks at sea and the regions on land that are in need of energy (cities, industrial areas). Second there is the incorporation of various supply technologies, like wind, wave, clean fossil fuel, biomass plants and even osmosis plants. It is highly likely that such a hybrid infrastructure is easier to install than a separate on-land grid-connection for each energy production unit. It is also possible to build extra storage capacity in the sea, like compressed air in underground salt domes or on a storage island, where water could be pumped up or down as it is needed.

With a combination of such local and centralized systems it is possible to fulfil energy demand at any time – just like we are now used to and how we would like to keep it. This is partly due to the storage potential in car batteries.

The method

In this chapter I have described a 'sustainable energy supply for everyone'. Within this term is contained a method. We start by building an inventory of our desires and needs and the will to fulfil them using as little external energy sources as possible. Then we determine how we can generate renewable energy in our own surroundings. And finally we make sure that we generate renewable energy where it is most appropriate and where it creates the highest yields.
Of course different options may be used. Everyone will develop his/her own sustainable energy system, depending on personal needs and the sources available. To do that we have to start thinking differently: creatively and out of the box. And especially: sustainably.

The egg has been boiled

A fresh outlook on energy

My journey towards 'a sustainable energy supply for everyone' started at the farm I grew up on. There I experienced the impact of the natural forces on daily life. The natural forces inspired awe, and justifiably so. These forces can form a threat. At the same time nature is a tremendous source of energy. The energy from the sun is even the source of our existence. For many millions of years the sun has given life on Earth the energy to develop. For many thousands of years humanity has used the sun as an energy source to satisfy its needs and desires.

Only in the last two centuries has humanity switched to using fossil fuels. Fossil fuels were easy to handle. Until today we prepare our meals on coal, gas, or oil, we drive in cars running on oil, we make wheat flour with the help of diesel, and our lights and appliances run on energy extracted from coal, gas, or oil. Fossil fuels were always present in abundance.

In the last century we discovered that fossil energy sources have their limitations. First we found out that heating or cooking on coal was unhealthy, and that electricity production with fossil fuels resulted in 'acid rain'. Luckily new technologies helped us keep this damage to a minimum.

Then the limited availability of fossil fuels was revealed. Sources that were once plentiful have dried up. It might take 50, 100, or maybe 300 years, but fossil fuels will eventually run out. We already have to make a considerably larger effort to mine fossil fuels than we did before. This happens at a higher cost and in locations that are difficult to reach. And it happens in beautiful places that were previously left untouched by human hand.

On top of this, fossil fuels are not fairly divided around the globe. And because everyone wants access to the sources that are available, sometimes even wars are being fought for them. When these sources decrease in quantity, tension between civilizations will increase.

And finally we are now uncovering the fact that the use of these fossil fuels is the main cause of climate change, an environmental problem on global scale. Despite all the technological, economical, and ecological improvements, fossil fuels are ultimately a dead end. There is however a route of escape: the sun, which will send us her rays for a few more billion years.

Sun and her derivates - wind, biomass, and waterpower – should not simply take the place of fossil fuels. We will have to develop an energy system that is fundamentally different from the current one. We have to radically change our way of thinking about energy.

We know that 98% of our energy is wasted by us. We also know that in one hour the sun provides the planet with so much energy that it equals our yearly energy consumption. Knowing this, we can build a new system. Sustainable energy is everywhere and freely available to anyone. To think of an efficient capture, storage and distribution system is the true challenge. It is difficult, but possible.

The character of sustainable energy demands a different design of the energy chain: from demand to supply, and not the other way around. We start locally and then move to power plants, not from power plants to the local area. We move from the consumer to the supplier, and not from energy company to energy users.
In this booklet I have shown that a fresh outlook on energy offers many opportunities. This starts with the sense of wonder at why things are the way they are. This new outlook opens our eyes about the enormous amounts of energy that are ready to be grasped.

We will need technology to fulfil our desires, but luckily most of these technologies already exist. Around the technologies we build our new systems. In this way a 'closed greenhouse', an air conditioning system that uses seawater, and a solar doorbell will go from design to widespread use.

This is not an ideology, nor am I moralising. A fresh outlook on energy offers countless opportunities for new companies, new jobs, new technology, and a new, clean environment. If companies and energy consumers make this outlook their own it will be a piece-of-cake to work together and build a sustainable energy supply for everyone.

The egg of Columbus

The new outlook on boiling an egg starts with the question: "what is an efficient way to boil an egg?" I have tried to find a solution. If we didn't have to use water we would make much progress, but boiling an egg in a pan without water is not going to happen. In the microwave we might be able to cook an egg. But at 1000 watts you might have to scrape the cooked egg off the sides of the microwave.

You can prick a little hole in the eggshell and try microwaving it at 600 watts. The egg is done after 45 seconds but without the shell the egg does not look particularly tasty. The egg white is firm, but the yolk is soft and puss-like. This is not the breakfast I had imagined.

Apparently there are special microwaves which in 8 minutes at 400 watts turn an egg into something I would consider edible. An egg-boiler on solar power exists as well, but unfortunately this is not appropriate for Dutch weather. There is too little sun to boil eggs year-round.

As long as we need water, the best way to boil eggs appears to be to put the eggs directly in the cold water and bring the water to a boil. Let it boil for one minute, and then switch off the stove, while leaving the egg in the hot water for 6 to 8 minutes. Reusing the cooking water is an obvious next step, and it can be done in the close-in boiler under the sink. With a second pipe in the sink, a heat exchanger, we can store the heat from the discarded water in the boiler. In theory this works, but I have not been able to find a plumber who can understand and install this ingenuous solution. And I couldn't do it myself.

There you are, with your fresh outlook on energy. On paper and in my head there are opportunities to spare, and none of them are too complicated. But the practical world is insubordinate.

There is some more time needed to sit with our ideas, especially to prepare the zeitgeist for the necessary adaptations. The developing and building of a sustainable energy system for everyone will also require everyone's cooperation. However, you and I, we will live to see it. The egg of Columbus - the sustainable egg - has been laid and now it needs to be boiled.

About the author

Prof. Dr. Ad van Wijk (1956) is one of the most influential sustainable energy entrepreneurs and innovators in Europe. In 1984 he co-founded the sustainable energy knowledge company Ecofys, which eventually grew into Econcern. As chairman of the executive board he led and developed Econcern into a company with 1200 people in more than 20 countries, sharing the mission 'A sustainable energy supply for everyone'. Until its bankruptcy in 2009 as a result of the financial crisis. Econcern was the holding company of Ecofys, Ecostream, Evelop, OneCarbon and Ecoventures.

Econcern developed and marketed many new products and services in the field of sustainable energy. Examples include the 120 MW offshore wind farm Princess Amalia, several multi-MW solar farms in Spain, BioMCN (a bio-methanol plant in Delfzijl) which is the largest second generation biomass plant in the world, Sea Water Air Conditioning systems in the Caribbean (still under development), the Closed Greenhouse (an energy producing greenhouse) the console (an innovative and cheap support structure for solar systems), the Energy Mirror (visualizing energy consumption in buildings) and Quicc (an electric van).

Ad van Wijk has been awarded many important prizes for excellent entrepreneurship. Amongst others he was entrepreneur of the year in the Netherlands in 2007. In that same year he received the 2007 Amsterdam Private Equity Club Award. And in 2008 he was top-executive of the year in the Netherlands. He was on several Supervisory Boards, amongst others Solland Solar Energy (a solar cell manufacturer) and BioMCN.

Today, Ad van Wijk is an independent sustainable energy entrepreneur, advisor and Professor in 'Future Energy Systems' at the Delft University of Technology. He is a member of the Economic Development Board Rotterdam and of the Royal Dutch Society of Sciences and Humanities.